THE OZONE HOLE

Sally Morgan

FRANKLIN WATTS
A Division of Grolier Publishing
NEW YORK • LONDON • HONG KONG • SYDNEY
DANBURY, CONNECTICUT

Picture Credits
Colorific!: page 29 left (Michael St. Maur Sheil). **Ecoscene:** pages 4 (Nick Hawkes), 9
(Wayne Lawler), 21 bottom (Rob Nicol). **Environmental Images/Photo Library:**
pages 12 (Mike Midgley), 15 top (Jim Miles), 26 (Steve Morgan). **Oxford Scientific
Films:** cover, small photo (Konrad Wothe) and pages 1 Ronald Toms), 6 top (Norbert
Rosing), 11 top (Kim Westerskov), 21 top (Warren Faidley). **Panos Pictures:** pages 14
(Jim Holmes), 29 right. **Planet Earth Pictures:** pages 5 bottom, 13 bottom (Space
Frontiers). **Science Photo Library, London:** pages 5 top (NASA), 10 (David Parker),
13 top (NASA), 15 bottom (Maximilian Stock Ltd), 17 (Simon Fraser), 20 (Simon
Fraser), 25 top (Simon Fraser). **Stock Market Photo Library:** cover main photo and
pages 6 bottom (Zefa/Dick Durrance II), 8, 11 bottom (Frank Rossotto), 16, 18
(Benjamin Rondel), 19 left, 19 right (J. Bator), 22-23, 23, 25 bottom, 27 top, 27 bottom,
28. **Telegraph Colour Library:** pages 22, 24.

Artwork: Raymond Turvey.

EARTH WATCH: THE OZONE HOLE was produced for Franklin Watts
by Bender Richardson White.
Project Editor: Lionel Bender
Text Editor: Jenny Vaughan
Designer: Ben White
Picture Researchers: Cathy Stastny and Daniela Marceddu
Media Conversion and Make-up: Mike Weintroub, MW Graphics
Cover Make-up: Mike Pilley, Pelican Graphics
Production Controller: Kim Richardson

For Franklin Watts:
Series Editor: Sarah Snashall
Art Director: Robert Walster
Cover Design: Jason Anscomb

First published in 1999 by Franklin Watts

First American edition 1999 by Franklin Watts
A Division of Grolier Publishing
90 Sherman Turnpike
Danbury, CT 06816

Visit Franklin Watts on the Internet at:
http://publishing.grolier.com

CONTENTS

THE OZONE HOLE

High above the earth there is an invisible layer of a gas called ozone. The ozone layer does a very important job. It stops invisible, harmful rays from the sun from reaching the earth's surface.

What Is Ozone?

Ozone is a gas naturally present in the environment. It is similar to the gas oxygen we breathe, but ozone has both a strong smell and a slightly blue color.

Ozone is easily damaged by chemicals used in industry. The chemicals break down the ozone, and this allows harmful rays from the sun to reach the earth. The ozone layer over the Antarctic region has become so thin that scientists say there is an "ozone hole."

We enjoy feeling the warmth of the sun's rays on our bodies. But sunlight contains rays that are harmful to living things. The ozone layer filters, or blocks out, these rays.

Solving the Problem

Scientists have been monitoring damage to the ozone layer for many years. They know how it is caused and what we can do to help correct the problem —we must reduce the amount of chemicals reaching the ozone layer.

The ozone layer is high above the clouds. In this view of the earth from space, the clouds are thickest over the Antarctic region (at the bottom of the picture). This is where the ozone layer is most seriously damaged.

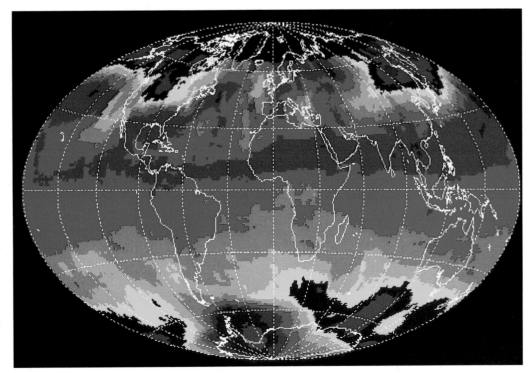

This computer image shows the amount of ozone the world over. Green colors show the highest levels of ozone, purple colors the lowest levels. Green areas turn purple as ozone is lost.

THE ATMOSPHERE

The earth is surrounded by air, which is a mix of gases known as the atmosphere. The atmosphere reaches up to about 430 miles (700 kilometers) above the earth, and merges into space.

The Layers of the Atmosphere

There are four layers in the atmosphere. The lowest is the troposphere. This contains about three-quarters of all the gases in the atmosphere, together with water vapor (tiny droplets of water) and dust. Clouds form in this layer.

The next layer is the stratosphere. This has warm and dry air. The ozone layer is in the stratosphere. The mesosphere lies beyond the stratosphere, and beyond that is the thickest layer, the thermosphere.

As the sun's rays warm the oceans, some liquid water turns to water vapor in the air. This helps to create clouds.

As these balloons rise up, they reach air that is colder and contains less oxygen.

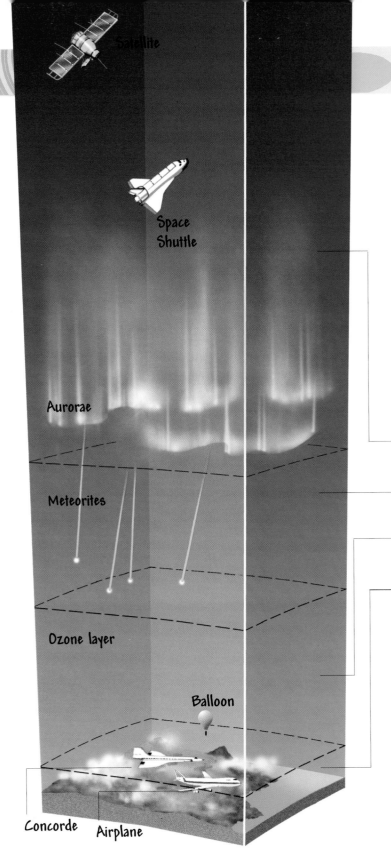

Satellite

Space
Shuttle

Aurorae

Meteorites

Ozone layer

Balloon

Concorde   Airplane

THE EARTH'S ATMOSPHERE
The different layers in the earth's atmosphere and objects
that may be seen in each layer.

THERMOSPHERE

MESOSPHERE

STRATOSPHERE

TROPOSPHERE

Taking Part

Green plants use carbon dioxide
and water to make their own food.
In the process, they release
oxygen. To see a plant produce
oxgyen, place some pondweed in a
bowl of water and set the bowl in
a sunny place. Soon you will see
tiny bubbles coming off the plants.
These bubbles contain oxygen.

Air

The air we breathe is part of the
troposphere. More than three-
quarters of this is made up of a
gas called nitrogen. About a fifth
of the air is the gas oxygen,
which almost all living things
need to survive. The rest is
made up of carbon dioxide gas,
water vapor, and tiny amounts of
other gases such as argon, neon,
and ozone.

THE OZONE LAYER

The ozone layer is like a thin slice of the stratosphere. It lies between 6 and 25 miles (10 and 40 kilometers) above the earth. That is the height at which supersonic aircraft such as Concorde fly. The ozone gas is spread out thinly and unevenly within this layer.

Filtering Ultraviolet

Sunlight contains invisible rays of light called ultraviolet light. These rays contain a lot of energy and can harm living things. The ozone layer filters out most of the ultraviolet rays as they enter the atmosphere, but it lets through warmth and visible light.

The sun's ultraviolet rays are what makes our skin tan. Too much ultraviolet can cause sunburn and wrinkling of the skin.

THE OZONE CYCLE
All chemicals are made of tiny building units called atoms. Atoms link together to form molecules.

Heat from the split ozone molecules goes into space.

1. Ultraviolet rays split ozone molecules

Oxygen atom

2. An oxygen molecule and an oxygen atom are formed.

3. The oxygen atom and oxygen molecule later combine and become ozone again.

Oxygen molecule

Good and Bad Ozone

Some chemical factories and office machinery, for example air conditioners and Xerox machines, produce ozone as a waste gas. This ozone stays in the lowest level of the atmosphere. There it helps absorb, or soak up, some of the ultraviolet rays that reach the surface of the earth.

Ozone itself is harmful, however. It irritates people's eyes, nose, and throat, and is especially harmful to people with breathing problems such as asthma and bronchitis.

Taking Part

Sunglasses block out some of the sun's ultraviolet rays. On a sunny day, stand outside and look around you. Can you see clearly in the sun? Now put on a pair of sunglasses. What difference do the sunglasses make? Can you see more or less with the glasses on? (CAREFUL! Never look directly at the sun.)

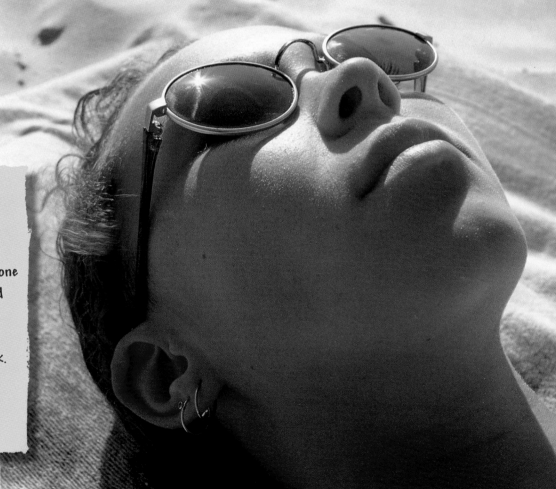

This woman is wearing sunglasses to prevent ultraviolet rays from damaging her eyes.

Eco Thought

If all the ozone molecules in the ozone layer were squashed together, the layer would be 1/4 inch (6 millimeters) thick. As it is, the ozone layer is several miles thick.

MEASURING OZONE

People have been studying the weather for hundreds of years. Since the 1960s, scientists all over the world have been studying the atmosphere as well.

A scientist is about to release a weather balloon into the lower atmosphere.

Weather Balloons

Scientists use weather balloons to find out what is happening in the lower atmosphere. The balloons are usually filled with helium, a gas that is lighter than air, producing a lifting effect.

Each balloon is attached to a piece of equipment called a radiosonde. The radiosonde collects information about the gases in the atmosphere. It sends this information back to earth using radio signals.

Measuring Ozone

At weather stations, the radio signals are fed into computers that are programmed to create maps showing the amounts of each gas in the atmosphere.

The first measurements of the ozone in the atmosphere were taken in 1956 at a research base in Halley Bay, Antarctica.

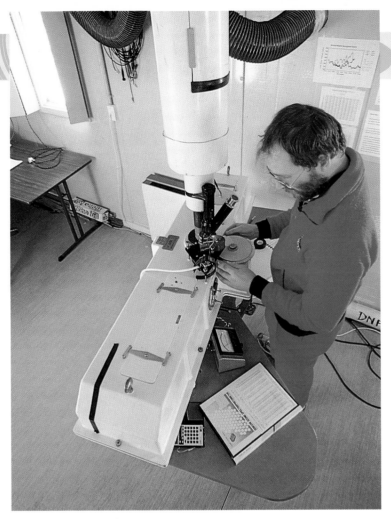

A scientist calculates levels of ozone in a sample of air taken from the atmosphere over Antarctica.

Using Satellites

Scientists use satellites to find out what is happening higher in the atmosphere. The first weather satellites were launched in the 1970s. The satellites orbit, or circle, the earth and measure air temperatures, moisture and sea levels, and wind speeds.

In 1978, the United States launched the Nimbus-7 satellite. This was the first satellite to measure how thickly the ozone molecules were spread in the ozone layer and how much ultraviolet light was being filtered out.

On the Ground

Nowadays, a weather balloon is sent up every day from the Halley Bay base in Antarctica. The measurements of ozone levels allow scientists to give early warnings of any sudden increases in ultraviolet radiation reaching the ground.

Skylab is a research station positioned in space. Like many weather satellites, it is launched into space by the Space Shuttle and sends back information about the earth's atmosphere.

DISCOVERING A HOLE

In 1985, a British scientist named Dr. Joseph Farman discovered a hole in the ozone layer over Antarctica. But the first warnings came earlier than that.

Eco Thought

Dr. Joseph Farman, the British scientist who discovered the Antarctic ozone hole, says that soon a third of the ozone layer over northern Europe and North America will be gone.

Early Warnings

In the early 1970s, scientists found that substances used in aerosol, or spray, cans damaged ozone molecules. The substances were used as propellants, making the spray-can's mechanism work. They consist of chemicals called CFCs (short for the technical name chlorofluorocarbons).

Many governments banned these chemicals. In 1985, scientists discovered that the damage was worse than they had thought. Levels above Antarctica were so low that they called the area a hole. They looked again at information from the Nimbus-7 satellite and realized that the hole had first appeared in 1976.

A scientist uses a Dobson machine in the Antarctic. This measures ozone levels in units called Dobson units.

This scientist is reading colored computer printouts that show the different levels of ozone found in the atmosphere as an aircraft flew over Antarctica.

Disappearing Ozone

Scientists are finding that the ozone layer is getting thinner everywhere in the Northern Hemisphere, rather than in just one place. It is getting particularly thin over the industrial regions of North America and Europe. This is where most of the ozone-damaging chemicals have been released. Ozone is being lost twice as fast now as it was ten years ago.

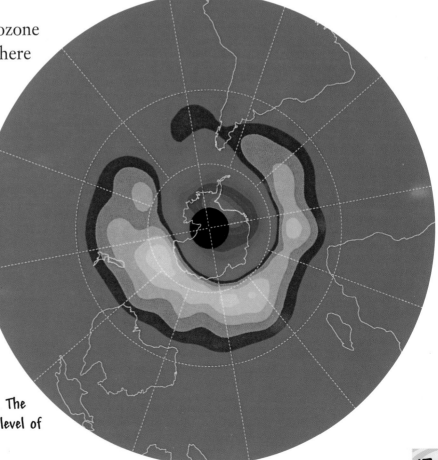

This computer image shows an ozone hole over the South Pole. The hole, where there is a very low level of ozone, is indicated in red.

CFCs and Ozone

Most of the chemicals that damage the ozone layer contain atoms of the gas chlorine. People are responsible for more than three-quarters of all the chlorine in the atmosphere. Most of it comes from CFCs, or chlorofluorocarbons. ('Chloro' in a chemical's name means it contains chlorine atoms.)

Wonder Chemicals

CFCs were invented in the 1920s. They have been used in aerosol cans, as cooling liquids in refrigerators and air-conditioning systems, for making the bubbles in polystyrene and other foams, and in making electronic components.

CFCs were cheap to make and are nontoxic (not poisonous). They seemed safe to use in industry. Their chemical structure does not break down easily, so they are long-lasting. Scientists in the 1920s called them wonder chemicals.

Mass Production

CFCs were so useful that they were made in large quantities and were used freely. The United States of America was the world's largest producer and exporter of CFCs. It was 50 years before scientists noticed the damaging effect of CFCs on the ozone layer.

Polystyrene fish boxes are light and strong and help keep the fish cold. Unfortunately, making them can release CFCs into the atmosphere.

Drifting Upward

The problems of CFCs are caused by the fact that they eventually release their chlorine. When they are released into the atmosphere, CFCs slowly drift upward into the stratosphere. They may take 10 years to reach the ozone layer, but they do not break down until they reach there. In the ozone layer, the chlorine atoms are freed. In certain conditions, chlorine atoms attack ozone molecules.

In the past, aerosol, or spray, cans like this contained CFCs that damaged the ozone layer. Modern aerosols work using harmless gases.

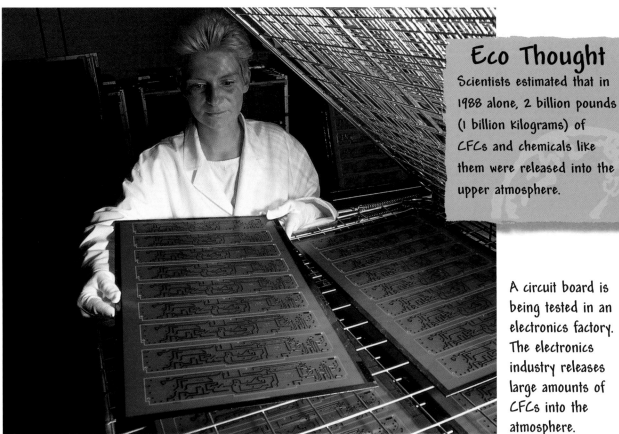

Eco Thought

Scientists estimated that in 1988 alone, 2 billion pounds (1 billion kilograms) of CFCs and chemicals like them were released into the upper atmosphere.

A circuit board is being tested in an electronics factory. The electronics industry releases large amounts of CFCs into the atmosphere.

DESTROYING OZONE

In the ozone layer, ozone is constantly being made and destroyed. Usually there is a balance between the amount that is made and the amount that is destroyed. CFCs, nitrogen oxides, and chlorine upset this balance so that more ozone is destroyed than is created. The imbalance creates ozone holes.

How Does Ozone Chemistry Work?

Oxygen molecules are made up of two oxygen atoms linked together. Ozone molecules are made up of three linked oxygen atoms. When ultraviolet light strikes oxygen molecules, the molecules split into two free oxygen atoms. One atom joins with an oxygen molecule to form an ozone molecule.

When the ozone molecules are struck by ultraviolet light, they break up. They form an oxygen molecule and one free oxygen atom. So, in the end, the amount of ozone does not change—unless other chemicals are involved in the process.

Suntan lotions help protect the skin from harmful ultraviolet rays.

Eco Thought

Some CFCs can survive 200 years, and so they have plenty to time to reach the stratosphere. When an atom of CFC reaches the ozone layer, it can destroy as many as 100,000 ozone molecules.

What Do CFCs Do?

In the stratosphere, CFCs are broken down by ultraviolet light, releasing chlorine atoms. These atoms attach themselves to ozone molecules. They help break down the ozone into separate oxygen molecules and atoms.

In this process, scientists call the chlorine a catalyst. A catalyst makes a chemical reaction occur more quickly and does not get used up in the process. So a single chlorine atom can be used again and again, breaking down many ozone molecules.

HOW CFCs DAMAGE THE OZONE LAYER

Fluorocarbon

Ultraviolet light from the sun

Heat released from splitting of ozone molecule

CFC molecule splits apart

Released chlorine atom

Ozone molecule

Oxygen molecule

breaks down

Oxygen atom

Chlorine atom —free to attack more ozone

Oxygen atom—joins with chlorine atom

This scientist in Hawaii is using special equipment for measuring the level of CFCs in the atmosphere.

17

OTHER ATTACKERS

CFCs are not the only industrial chemicals attacking the ozone layer. Waste chemicals from burning fuels, pesticides, and cleaning fluids also damage ozone.

Aircraft Engines

One group of ozone-attacking chemicals are nitrogen oxides. They are released by jet aircraft engines as they burn fuel in the stratosphere. The nitrogen oxides speed up the breakdown of ozone. But they do not get used up themselves in the chemical process. This means that the nitrogen oxides can damage the ozone over and over again.

Eco Thought

An estimated 3.5 million tons of nitrogen oxides are released by passenger jet aircraft every year as they fly close to the ozone layer.

The Space Shuttle

The rocket boosters of the U. S. space shuttle release the gas hydrogen chloride when the vehicle is being launched. This is an ozone-attacking gas that is released naturally from volcanic eruptions. In the stratosphere, the chlorine atoms in the hydrogen chloride split ozone molecules into oxygen molecules and atoms.

Concorde supersonic aircraft release ozone—damaging gases directly into the stratosphere.

As the space shuttle is launched, hydrogen chloride gas is released and drifts toward the ozone layer.

Small aircraft with liquid sprays are often used to spread pesticides over large areas. The spray lingers in the air.

Eco Thought

Each shuttle launch produces about 68 tons of hydrogen chloride. Most is released in the troposphere, the lowest part of the atmosphere.

Pesticides and Cleaning Agents

Farmers spray pesticides on their crops to kill pests such as insects and fungi. Some of these sprays contain a chemical called methyl bromide. The bromide part of this chemical attacks ozone. (Bromides contain bromine, a gas similar to chlorine and fluorine.) Methyl bromide destroys ozone 40 times more powerfully than CFCs.

Dry cleaning uses chemicals called solvents to remove dirt from clothes without making them wet. Some of the chemicals release chlorine, further damaging the ozone layer.

NATURAL CAUSES

Some of the chlorine that damages the ozone layer comes from natural sources, such as volcanic eruptions and forest fires. We cannot control when or how much of this chlorine is produced.

Erupting Volcanoes

When volcanoes erupt, they release hundreds of tons of water vapor and chemicals into the atmosphere. The chemicals include hydrogen chloride, which is a mixture of hydrogen and chlorine atoms. The more powerful eruptions send a cloud of water vapor and chemicals beyond the troposphere and into the stratosphere.

An erupting volcano in Hawaii pours chemicals into the atmosphere.

Up in the Troposphere

Scientists studied in detail the eruptions at El Chichon in Mexico in 1982 and Mount Pinatubo in the Philippines in 1991. Using instruments on weather balloons, aircraft, and satellites, they measured the makeup of volcanic clouds as they rose in the atmosphere.

The scientists discovered that ice in the troposphere removed most hydrogen chloride from the volcanic clouds, but a small amount reached the ozone layer. The resulting thinning of the ozone layer lasted a few years.

Eco Thought

Scientists think that when Mount Pinatubo erupted in 1991, this may have helped increase the size of the ozone hole over Antarctica. By the spring of 1992, the ozone hole had grown by one-fifth.

Sparks in the Sky

Lightning is enormous sparks of electricity between clouds or between clouds and the ground. When lightning occurs high in the sky, it can set off reactions that split ozone molecules, thinning the ozone layer.

Burning Forests

A burning forest releases many different gases. Some contain chlorine. The gases rise into the stratosphere and affect the ozone hole. Huge fires of the last few years will have released enormous amounts of ozone-damaging gases.

As lightning flashes through the sky, it both creates ozone from oxygen and makes some chemicals in the air attack ozone.

As forest fires burn, they release gases into the air. These gases can drift upward and damage the ozone layer.

INSIDE THE HOLE

Each year in Antarctica, the ozone hole appears toward the end of winter, in August. The hole gets bigger during spring, in September and October. By the end of October, the levels of ozone are less than half the normal level.

Cut Off From Other Regions

In winter, strong winds blow around the Antarctic, cutting off the air there from the rest of the world. At that time of year, the sun lies below the horizon and air temperatures can fall to –112°F (–80°C). The water in the air becomes ice.

These conditions are right for chlorine atoms to attack ozone molecules. In spring, the sun rises in the sky and its ultraviolet rays split CFCs, releasing more chlorine atoms. They destroy most of the ozone between 9 and 12 miles (14 and 20 kilometers) above the ground.

Clouds form as water vapor in the air cools and condenses into water droplets. In cold temperatures, the droplets form ice crystals.

Ice crystals like this are formed by water freezing in the atmosphere. The ice crystals speed up the breakdown of ozone molecules.

The Hole Breaks Up

In November, winds carry warmer ozone-rich air into the Antarctic, and the hole is repaired. But air containing low levels of ozone drifts north over New Zealand, Australia, and South America. In these places, extra high levels of ultraviolet light reach the ground.

A scientist studies how much ultraviolet light gets through the ozone layer and reaches crops.

A Hole Over the Arctic

The weather conditions in the Arctic are different from those in the Antarctic. Instead of forming a hole, the ozone layer is getting thinner all over. The lowest levels of ozone occur in the Arctic winter, in February.

THE EFFECTS ON US

Our skin needs to receive some ultraviolet light to help produce vitamin D, a chemical needed for healthy bones. But too much ultraviolet light harms our bodies in many ways.

Sensitive Eyes

Our eyes are very sensitive to ultraviolet rays. This is why we should wear sunglasses in bright light. Someone whose eyes have been damaged by ultraviolet rays may seem to see snow falling. The condition is called snow blindness. Sometimes the eye clouds over, causing blindness. This is called a cataract.

An optician examines a girl's eyes. Too much ultraviolet light can lead to sight problems.

Eco Thought

The United States Environmental Protection Agency (EPA) estimates that if ozone continues to be destroyed at the same rate as now, each year another 50,000 people could suffer from skin cancer and an additional 100,000 people become blind as a result of ultraviolet damage to their eyes.

Cancer Cells

People's skin cells absorb the ultraviolet rays. The first sign of damage to the skin is burning. If this happens often, and if ultraviolet light reaches the skin for long periods of time, skin cells can develop into cancer cells. These grow out of control and divide over and over again.

Skin cancer and eye problems are becoming more common in southern countries such as Australia and Chile. They affect people and animals such as cattle and sheep, too.

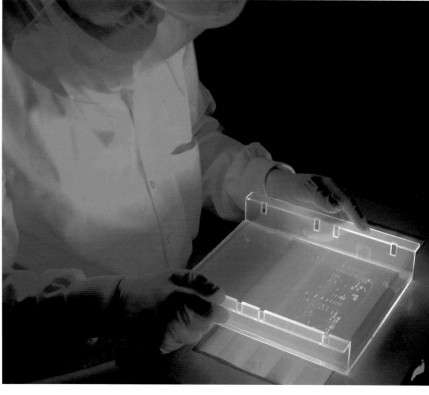

A researcher uses ultraviolet light to examine DNA, a chemical in body cells. DNA is damaged by high levels of ultraviolet light.

A scientist studies human body cells that have been exposed to ultraviolet light.

On the Ground

The "slip, slap, slop" campaign in Australia reminds people about the dangers of being out in the sun. People are encouraged to slip on some clothes to cover their arms and legs, slap on a sun hat, and slop on some sun cream on any bare skin.

REPLACING CFCs

The only way to stop the damage that is being done to the ozone layer by the CFCs is to stop using them. That means that we must find safe ways of doing the jobs CFCs used to do.

If fewer CFCs are released and the ozone layer starts to grow thicker again, the first place it will be noticed is here in the Antarctic.

International Agreements

In 1987, 31 nations signed the Montreal Protocol. This was an agreement to cut by half the use of CFCs by the year 1999. In 1992, 100 nations agreed to stop producing CFCs and chemicals like them by the year 2000. The European Union and the United States agreed to stop using these substances by 1995. By 2030, the world may be free of these ozone-attacking chemicals.

On the Ground

Only in 1998 did scientists find a replacement for the CFCs used in the special inhalers needed by people who suffer from asthma. The inhalers help the people to breathe more easily. Soon, all asthma inhalers will be CFC-free.

Using Alternatives

Industries that produce foam fillings, air-conditioning units, polystyrene, or refrigerators now use two other artificial chemicals instead of CFCs. Both replacements still damage the ozone layer, but not as much as CFCs.

There are some natural substances that could replace CFCs and that do no damage at all. The best ones are called hydrocarbons. These are substances that contain carbon, a chemical commonly found in wood and coal. They include the gases propane and butane. They are just as cheap as CFCs and can do some jobs better.

New refrigerators that use non-CFC coolants are checked for safety in a warehouse.

A technician assembles an air-conditioning unit that uses an ozone friendly chemical instead of CFCs.

Eco Thought

The levels of CFCs in the atmosphere are still increasing. This is because the CFCs released in the 1980s and 1990s have just reached the ozone layer. The levels are expected to be highest soon. If nations can keep to the agreements they have signed, ozone levels should return to normal by 2045.

WHAT CAN WE DO?

There are still lots of CFCs around. If you look around your home, you may find old aerosol sprays containing CFCs. Old refrigerators, freezers, and fire extinguishers all contain CFCs or related chemicals.

Aerosol Cans

We still use aerosol cans, but now they all have a sign on them to say they are ozone friendly. The CFC propellant has been replaced by one that does not damage the ozone.

Pump-action sprays are even better. They do not contain any propellant. Instead, you have to pump air into the can, and the contents squirts out. If you have a choice, buy this type of spray.

This is a special site for toxic waste in the United States. Harmful chemicals such as CFCs can be dumped here safely.

The Ozone Friendly Refrigerator

Is your family planning to get a new refrigerator? If your old refrigerator uses CFCs, get these chemicals disposed of safely. Then buy one of the latest designs of refrigerator that are ozone friendly and use less than half the energy as normal ones. The walls of the machine are insulated with a foam that has bubbles of hydrocarbons instead of CFCs.

A street-vendor in Thailand sells second-hand electronic components. Making components produces CFCs, so recycling helps reduce them.

This engineer is draining the CFC coolant from an old refrigerator in order to dispose of the chemicals safely.

On the Ground

Ghana imports a lot of second-hand refrigerators from Europe. These still contain CFCs. Some refrigerator manufacturers are training local people to convert these to use propane gas, to make them ozone friendly.

FACT FILE

Smelly Gas

The word "ozone" comes from a Greek word that means "smell." This is because ozone has a very pungent (strong) smell. Oxygen, a similar gas, has no smell.

A Ban on CFCs

In 1979, the United States and several other countries completely banned the sale of aerosol cans containing CFCs. Since January 1996, it has been illegal for CFCs to be manufactured in the United States.

Sweden First

In 1988, Sweden became the first nation in the world to pass laws to completely remove CFCs from all household equipment and industries in the country. It achieved this in 1994.

In the Year 2030

In November 1992, a new agreement to get rid of CFCs was signed by more than 100 nations. This covered more than nine-tenths of the world's CFC users. The nations agreed that they would phase out CFCs by the year 2030. They would also get rid of halons, which are ozone-attacking chemicals like CFCs and which are used in many types of fire extinguishers.

Worse Than Expected

In April 1991, American scientists announced that ozone destruction over the highly populated regions of the world was two to three times greater that they had previously thought. This was particularly true for North America and Europe.

Different Types of Ultraviolet Light

There are three forms of ultraviolet rays. Scientists refer to these as UV-A, UV-B, and UV-C. UV-A is the least damaging to living things, while UV-C is the most damaging. Fortunately, the ozone layer filters out all of the UV-C. The amount of UV-B reaching the ground is least where sunlight has to pass through the most atmosphere. A person standing in sunlight at the equator would receive 1,000 times more UV-B than a person standing at the North or South Pole.

Measuring Ozone

Levels of ozone are measured in Dobson units. One Dobson unit is about 27 million molecules of ozone per 0.15 square inch (1 sq. centimeter).

GLOSSARY

Aerosol A fine spray. Aerosol cans spray out the liquid inside them in a fine mist.

Atmosphere The mix of gases surrounding the earth. The atmosphere is made of four main layers.

Atom The tiniest possible part of any substance. Atoms are invisible to the naked eye.

Cancer A disease where cells in the body grow out of control. Cancer cells divide over and over again, producing a large mass known as a tumor.

Catalyst A substance that speeds up a chemical reaction but does not take part in the reaction itself.

Cataract A disease of the eye where the lens becomes cloudy, causing blurred vision.

CFC Chlorofluorocarbon, an artificial gas once used in refrigerators and aerosols. It is a mixture of chlorine, fluorine, and carbon atoms.

Condense A change in the state of a substance, from gas to liquid or from liquid to solid, for example. As steam from a kettle cools, the water vapor condenses to liquid water.

Coolant A liquid, such as water, that removes heat from an engine or machine.

Hydrocarbon Chemicals that contain atoms of hydrogen and the substance carbon, for example, oil and natural gas.

Mesosphere The third highest layer of the atmosphere.

Molecule A group of two or more atoms linked together. The atoms can be of identical or different chemicals.

Nontoxic Not poisonous or harmful.

Pesticide A substance that will kill insect pests such as mosquitoes and aphids.

Polystyrene A lightweight but rigid (stiff) white substance used for insulation and packaging.

Propellant A gas used in an aerosol can to help the spray mechanism work.

Radiosonde A balloon for measuring weather conditions high in the atmosphere.

Satellite An instrument that is placed in orbit around the earth or other body in space.

Smog A thick, yellow haze over cities, created by pollution of the air from industry and cars.

Solvent Anything that makes another substance dissolve, or go into solution; for example, hot water dissolves sugar.

Stratosphere The second highest layer of the atmosphere. The ozone layer lies within the stratosphere.

Supersonic Traveling faster than sound, which travels at about 770 miles (1,240 kilometers) an hour in air at ground level.

Thermosphere The highest layer of the atmosphere.

Troposphere The lowest layer of the atmosphere. It contains the air we breathe in and out.

Ultraviolet An invisible part of sunlight that can harm the human body.

Volcano A gap in the earth's crust where hot rocks and lava may spill out onto the earth's surface and dust and gases may be shot up into the upper layers of the atmosphere.

INDEX